Bibliographic information published by the German National Library:

The German National Library lists this publication in the National Bibliography; detailed bibliographic data are available on the Internet at http://dnb.dnb.de .

Imprint:

Print and binding: Books on Demand GmbH, Norderstedt Germany
ISBN: 9783668447486

This book at GRIN:

http://www.grin.com/en/e-book/358591/design-considerations-for-ground-clearance-and-wheel-base-for-navigating

Venkatnarayan Hariharan

Design considerations for ground-clearance and wheel-base for navigating non-standard speed-bumps

GRIN Publishing

DESIGN CONSIDERATIONS FOR GROUND-CLEARANCE AND WHEEL-BASE FOR NAVIGATING NON-STANDARD SPEED-BUMPS

Venkatnarayan Hariharan

ABSTRACT

In some countries, there is very little standardization in the construction of speed-bumps (aka. speed-breakers), as a result of which the bottom of many cars scrape the speed-bumps. In this paper, we come up with a first order figure of merit that car manufacturers can use when adapting their cars to be sold in such countries, so as to minimize the chances of scraping the speed-bumps.

Keywords: Bottom, Car, Ground-clearance, Scrape, Speed-bump, Wheel-base

Nomenclature

`L` : Length of speed-bump, mm

`h` : Peak height of speed-bump, mm

`g` : Ground-clearance of vehicle, mm

`w` : Wheel-base of vehicle, mm

`r` : Radius of the wheels, mm

1. Introduction

In some developing countries, there is very little standardization in the construction of speed-bumps (aka. speed-breakers). The height and profile of the speed-bumps are very non-standard, as a result of which the bottom of many cars scrape the speed-bumps. This is a safety concern because such repeated impacts to a monocoque structure can cause damage to the body shell, besides resulting in an unpleasant driving experience. This is especially a problem for international car manufacturers who wish to sell their cars in that country. In this paper, we come up with a first order model for the figure of merit that car manufacturers can use when adapting their cars to be sold in such countries, so as to minimize the chances of scraping the speed-bumps. Car buyers can also use this model to evaluate one car versus another, as regards susceptibility to scrape speed-bumps.

Contrary to conventional belief, ground-clearance alone is not a criterion for scrape-susceptibility. The wheel-base also plays a role, especially for long speed-bumps. Indeed, if the length of the speed-bump is longer than the vehicle's wheel-base, then the ground-clearance requirements of the vehicle can often be relaxed, which is a desirable goal for manufacturers to get better vehicle dynamics from a stability and ride/handling point of view.

In this paper, we develop a model that shows the inter-play between the ground-clearance and the wheel-base. We develop the model in section 2, and we present the results in Section 3.

2. Model

Non-standard speed-bumps come in varying profiles, as seen in Fig. 1. It is very difficult to model scrape-susceptibility for all the various profiles. Hence, as a first order model, we develop the model for a simple isosceles triangular profile shown in Fig. 1b.

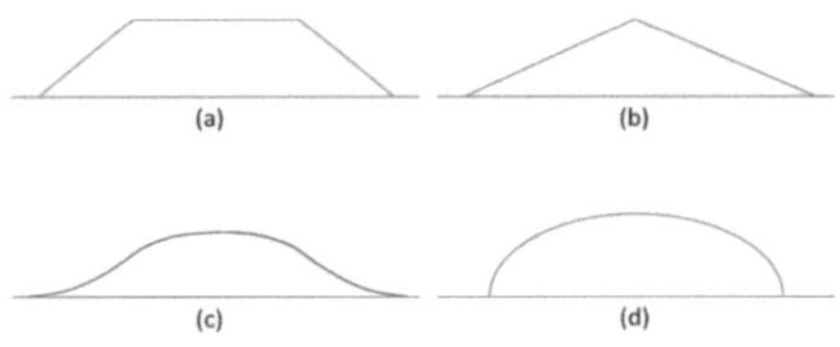

Fig. 1: Profiles of various speed-bumps.

Consider a vehicle negotiating such a speed-bump. It is clear that if the speed-bump is short enough ($L < w$), then the ground-clearance has to be greater than the speed-bump's height ($g > h$). This is the trivial case. But most often, speed-bumps are longer and have a smoother up-ramp and down-ramp. Because of this, when the front wheels have crossed the peak of the speed-bump but before they completely cross the speed-bump, the rear wheels start climbing the up-ramp. This raises the height of the lowest point of the vehicle's bottom, and hence reduces the chances of scraping the speed-bump. This is the reason why the ground-clearance requirements can be relaxed when the wheel-base is shorter than the speed-bump's length. This is the non-trivial case, and this is what we model in the next section.

2.1. Model Formulation

Fig. 2 shows the situation described above, at the point when the rear wheel is just about to start the up-ramp and the bottom is about to scrape the speed-bump.

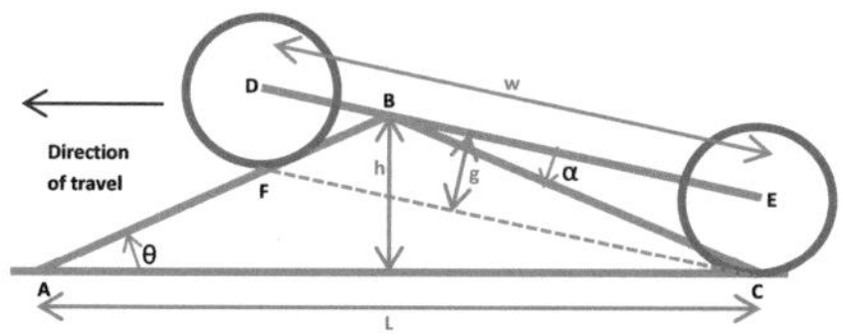

Fig. 2: Vehicle navigating a speed-bump, traveling right to left.

By formulating a model for this condition, we would have essentially formulated a condition to prevent scraping. It should be noted however that this would be a *necessary* condition and not a *sufficient* condition to prevent scraping. As the rear wheels climb the up-ramp, the front wheels descend the down-ramp. Depending on whether the descent is greater or the ascent is greater, a sufficient condition may be derived. But that is left for a future work.

Consider Fig. 3 which is a detailed view of the rear wheel. Line DE represents the bottom of the vehicle. Using elementary geometric relations, we can see that:

$$2h = L \tan\theta \qquad (1)$$

Using the SINE rule in triangle PEO, we can see that:

$$\frac{g-r}{\sin\left(\frac{\pi}{2}-\alpha\right)}=\frac{OP}{\sin\left(\frac{\pi}{2}+\alpha-\theta\right)} \quad (2)$$

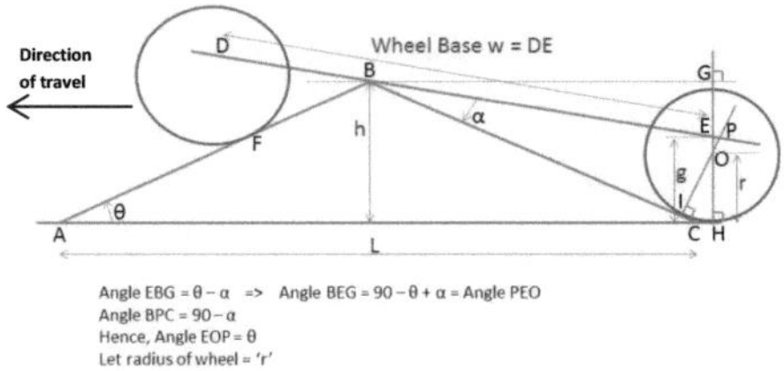

Fig. 3: Analysis of the rear wheels (right end of figure)

In triangle BPC, we can see that:

$$\tan\alpha=\frac{PI}{BI}\cong\frac{PI}{BC}=\frac{OP+r}{\sqrt{h^2+L^2/4}} \quad (3)$$

Eliminating OP in (2) and (3), we get:

$$2rcos\alpha+2(g-r)cos(\theta-\alpha)=\sqrt{L^2+4h^2}tan\alpha \quad (4)$$

Let us analyze the situation at the front wheel now. Consider Fig. 4 which is a detailed view of the front wheel.

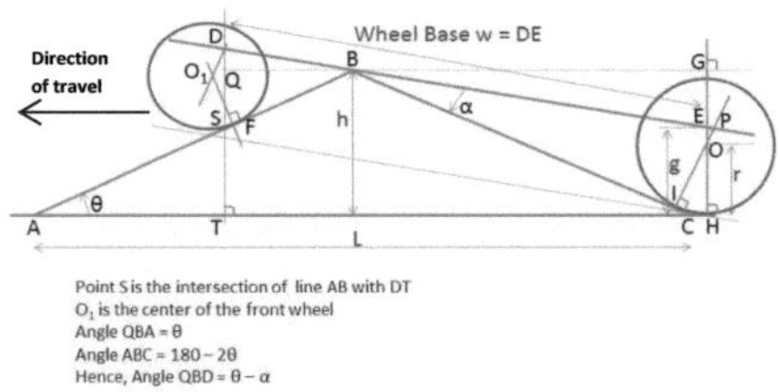

Fig. 4. Analysis of the front wheels (left end of figure)

In triangle BQD, we can see that:

$$QD = BD \cdot \sin(\theta-\alpha) \quad (5)$$
$$QB = BD \cdot \cos(\theta-\alpha) \quad (6)$$

Also:

$$QS = DS - QD \cong g - QD \quad (7)$$
$$QS/QB = \tan(Angle\ QBA) = \tan\theta \quad (8)$$

From (5 - 8), we have:

$$\tan\theta = \frac{g - BD \cdot \sin(\theta - \alpha)}{BD \cdot \cos(\theta - \alpha)} \qquad (9)$$

But $BD = w - BE \ \cong w - g/sin\alpha$

Hence we have:

$$\tan\theta = \frac{g - (w - g/sin\alpha) \cdot \sin(\theta - \alpha)}{(w - g/sin\alpha) \cdot \cos(\theta - \alpha)} \qquad (10)$$

Since we are considering realistic speed bump dimensions in the non-trivial case of $L > w$, we can approximate $\frac{h}{L} \to 0$ (h will typically be of the order of g, and we can see from realistic dimensions in Table 1 that this is a very reasonable approximation). We can see from (1) that this then implies that $\theta \to 0$. Then, since $\alpha \leq \theta$, this further implies that $\alpha \to 0$.

Hence $\sin\alpha \cong \alpha \cong \tan\alpha$, $\sin(\theta - \alpha) \cong \theta - \alpha$, $\cos(\theta - \alpha) \cong 1$, and $\tan\theta \cong \theta$.

With these approximations,

(1) becomes:

$$2h \cong L\theta \qquad (11)$$

(4) becomes:

$$2g \cong \alpha\sqrt{L^2 + 4h^2} \cong \alpha L \qquad (12)$$

and (10) becomes:

$$\theta \cong \frac{g - (w - g/\alpha) \cdot (\theta - \alpha)}{(w - g/\alpha)} \qquad (13)$$

Eliminating α and θ in equations (11), (12) and (13), we get:

$$L = w\left(2 - \frac{g}{h}\right) = L_{FOM} \qquad (14)$$

Equation (14) suggests a figure of merit that buyers can use when comparing different vehicles, or that manufacturers can use in finalizing the dimensions when there is some room to do so after other design considerations. From a scrape-susceptibility point of view, the smaller the value of L_{FOM}, the better. In other words, it is desirable that the vehicle can accommodate as small a speed-bump as possible (ie. small in length L; not to confuse with short in height h).

It is interesting to note that L_{FOM} is independent of the radius of the wheel r. This is quite reasonable since it is a necessary condition and hence depends directly on the ground clearance. It seems reasonable to expect that the radius of the wheel will play a role in modeling the sufficient condition, where rates of ascent and descent would come in the picture.

3. Results

Table 1 shows the ground-clearance and wheel-base dimensions of various cars sold in India [1 – 7]. Fig. 5 shows a plot of L_{FOM} for those cars. When deciding which car to purchase, all other things being equal, the buyer can go with the car having the smallest value of L_{FOM}.

Table 1: Ground-clearance and wheel-base of various cars sold in India (from [1 - 7]).

	GC (mm)	WB (mm)
Maruti Suzuki SX4 VXi	170	2500
Hyundai Verna Fluidic	170	2570
Volkswagen Vento	168	2552
Honda City	165	2600

Hyundai i20	165	2525
Hyundai i10	165	2380

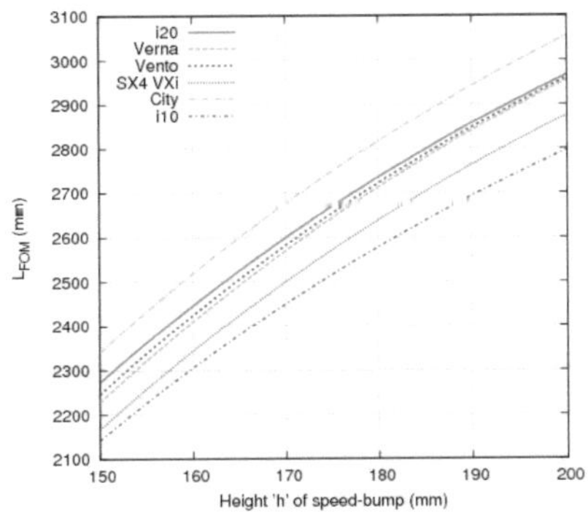

Fig. 5. L_{FOM} plots for various cars.

From Fig. 5, the Hyundai i10 has the best immunity from scrapes as it has the lowest L_{FOM}. This is the case inspite of it having the lowest GC. This is attributed to the fact that it has the smallest WB. In the sedan category, the Maruti Suzuki SX4 has the best immunity and the Honda City has the worst. This is indeed verified from driving experiences in various cities in India where various drivers have reported that the Honda City's underbelly scrapes most speed-bumps whereas the SX4 runs unscathed. The i20, Vento and Verna are closely bunched together, with their differences narrowing for taller speed bumps. It should be noted that this is the case even though the i20 has a lower GC than the other two. Again, this is attributed to the fact that it has a smaller WB.

4. Conclusion

An analytical model has been presented that vehicle manufacturers can use in competively designing the ground-clearance and wheel-base of their vehicles for safely negotiating non-standard speed-bumps. It has also been shown how potential car buyers can quantitatively use the dimensions data from brochures in arriving at a figure of merit (FOM), when deciding which car is safer for negotiating speed-bumps, assuming all their other purchase criteria have been met.

REFERENCES

1. Maruti Suzuki SX4 specifications from manufacturer (2014). http://www.marutisuzuki.com/sx4.aspx#dime

2. Hyundai Verna specifications from a car portal (2014). http://www.carwale.com/hyundai-cars/verna/fluidic14vtvtgl-specifications-3312.html

3. Volkwagen Vento specifications from manufacturer (2014). http://www.volkswagenlive.in/volkswagenform/ebrochures/vento_ebrochure.pdf

4. Honda City specifications from manufacturer (2014). https://www.hondacarindia.com/AllNewCity/specifications.html

5. Hyundai i20 specifications from manufacturer (2012). http://www.hyundai.com/wcm/upload_wwn/IN/ebrochure/2012_i20_ebrochure_1388197900544.pdf

6. Hyundai i20 specifications from a car portal (2014). http://www.carwale.com/hyundai-cars/i20/era12-specifications-2491.html

7. Hyundai i10 specifications from manufacturer (2011). http://www.hyundai.com/wcm/upload_wwn/IN/ebrochure/2011_i10_ebrochure_1386223196921.pdf